Fireworks

WHERE'S THE SCIENCE HERE?

Zambelli's rainbow and points of light shells with tiger tails over the city of Pittsburgh

Fireworks
WHERE'S THE SCIENCE HERE?

VICKI COBB

Photographs by
Michael Gold

 Lerner Publications Company   Minneapolis

For my grandson, Ben Trachtenberg, who has a history of watching fireworks with us.

Acknowledgments:
The author extends greatful appreciation to Philip Butler and Donna Grucci Butler and the staff of Fireworks by Grucci for showing how fireworks are constructed, and to Marcy Zambelli and Dr. George Zambelli for reviewing the final text and photographs, but she accepts full responsibility for the accuracy of the text.

Cover photograph courtesy of SuperStock, Inc./ SuperStock. All interior photographs courtesy of Michael Gold except the following: © Michael Freeman/CORBIS: p. 15; © Gunter Marx Photography/CORBIS: p. 19; © David Butow/ CORBIS SABA: p. 21; © Reuters/CORBIS: p. 24

Lerner Publications Company
A division of Lerner Publishing Group
241 First Avenue North
Minneapolis, Minnesota U.S.A.

Website address: www.lernerbooks.com

Library of Congress Cataloging-in-Publication Data
Cobb, Vicki.
Fireworks / by Vicki Cobb.
p. cm. — (Where's the science here?)
ISBN-13: 978-0-7613-2771-4 (lib. bdg. : alk. paper)
ISBN-10: 0-7613-2771-1 (lib. bdg. : alk. paper)
1. Explosives—Juvenile literature.
2. Fireworks—Juvenile literature. I. Title.
TP270.5.C63   2006   662'.1—dc22
2004029823

Manufactured in the United States of America
2 3 4 5 6 7 – DP – 11 10 09 08 07 06

Contents

A stationary ground display is studded with rolled, thin, cylindrical papers containing pyrotechnic powder. When it is ignited, designs, pictures, and portraits light up in different colors.

You'll Get a Bang Out of This

What's your favorite kind of firework? Is it a white chrysanthemum leaving tracers in the sky? Is it a golden flitter that leaves a trail of burning embers? A blue and gold spider or two soaring palms? Maybe you get the biggest bang out of a flash of light and a punch-in-the-stomach boom. Just about everyone loves the grand finale—a barrage of sound and light with Roman candles, tiger tails, and gold-spangled chrysanthemums.

Painting the sky with light and sound is the job of *pyrotechnicians*—people specially trained to set up and set off fireworks displays. It is not a job for amateurs. Fireworks are dangerous and can be deadly.

Red and silver chrysanthemums and red bombette candle fans light the Pittsburgh skyline.

A shell is launched from a mortar rack on its way to an aerial explosion.

The invention of modern fireworks involved playing with fire, often with disastrous results. But, like moths to a flame, people found them irresistible, and so they came to be.

What are fireworks? What makes them explode? How are the colors created? How can an explosion be timed for the right moment? How can one shell produce many different effects? These are questions that are answered by science. Fireworks displays might be considered show business, but fireworks themselves are a product of chemistry and physics. The ideas of the science behind fireworks are pretty amazing, too. This book tells the story.

Playing with Fire

You don't need a match to make a fire. All you need are lightning and some dry wood. That's probably how our ancestors discovered this very useful but also dangerous source of heat and light. At first, since no one knew how to start a fire, people probably spent a lot of time keeping a fire going. It was easier to add fuel than to wait for the next lightning strike.

In time, people discovered ways to start fires. Striking a certain rock with a piece of metal could make sparks. The sparks could ignite tiny pieces of fuel, called tinder. On a sunny day a glass lens could focus the heat of sunlight on tinder so that it burned. Or the heat made by twirling the end of a stick into a piece of wood would ignite it.

A Boy Scout starts a fire by striking a "flint" stone against a piece of steel.

Another primitive method of starting a fire is by using a wooden bow and drill. The bow spins a pointed stick in a shallow hole. This produces some hot, glowing sawdust that is quickly dumped onto a nest of tinder to produce a flame.

Once made, fires could warm homes, cook food, light the darkness. Fire could transform rocks into metals, such as iron and bronze, and melt sand into glass. Fire became so important to our ancestors that they thought of it as one of the four basic elements: earth, water, air, and fire.

People knew that it took three things to make a fire: fuel, air, and heat to get it started. But for centuries they didn't know what fire really is. In the middle of the nineteenth century scientists called chemists looked upon a flame as a puzzle to be solved. After many experiments they learned that a flame is produced as part of something called a "chemical reaction."

In chemical reactions, elements and compounds combine with each other. In the process, energy can be given off or taken up. Fire is an especially rapid chemical reaction called *combustion*.

The Element of an Element

The idea of four basic elements was replaced with the idea of a chemical element. An element is something that cannot be broken down into a simpler substance. There are ninety-two elements in nature (and more that have been made in the laboratory). They include oxygen, hydrogen, gold, carbon, sulfur, and nitrogen. Elements combine in chemical reactions to form compounds. A compound is a substance composed of two or more elements, and it is often very different from its combining elements. Water is one of the most common compounds. It is made of the elements hydrogen and oxygen, which are both gases. Carbon dioxide is another common compound. It is a gas made of carbon (a black solid) and oxygen.

In combustion, a fuel such as wood or wax combines with oxygen in the air to form water vapor and carbon dioxide. In the process, a lot of heat and light energy is given off. The light of a flame comes from glowing bits of very hot carbon (from the fuel) as they combine with oxygen.

A spark or a source of heat is needed to get combustion started. But once a fire is under way, the heat energy given off by the reaction itself is more than enough to keep it going. Modern chemistry began when chemists understood the chemical reaction that produces a flame.

Once a fire is roaring, new logs can be added. The fire is now hot enough for them to burn.

The Anatomy of a Flame

Here's an experiment in which you can take apart a flame as the early chemists did. You will need a candle in a candleholder, a match, a white china plate, and an adult helper, since you will be using fire.

Ask an adult to light the candle. When it is burning well, notice the bluish color at the bottom of the flame. That is where hydrogen from the fuel is combining with oxygen. Hold the plate over the yellow part of the flame for a few seconds. Look at the plate. The soot on it comes from glowing carbon that was cooled down by the plate. If you look closely you may see tiny drops of water that condensed on the plate. The water comes from hydrogen combining with oxygen from the air. When carbon combines with oxygen, carbon dioxide—the other product of combustion—is formed. Carbon dioxide is a colorless and odorless gas and is even harder to detect than water vapor.

Black Powder

Black powder is one of the most important ingredients of fireworks. But its use in fireworks wasn't as important as its use as gunpowder and in explosives for construction. For centuries, people made lots of money producing black powder for armies. As you might expect, there is some science behind the explosive nature of black powder.

Caves containing bat droppings and piles of animal manure sometimes have a white crust of powdery crystals. When these crystals are sprinkled on a fire, the fire burns more brightly. The crystals are an amazing substance

Black powder was important for hunters and soldiers in the eighteenth century. They kept their powder dry in leather bags or the hollow horns of cows or oxen.

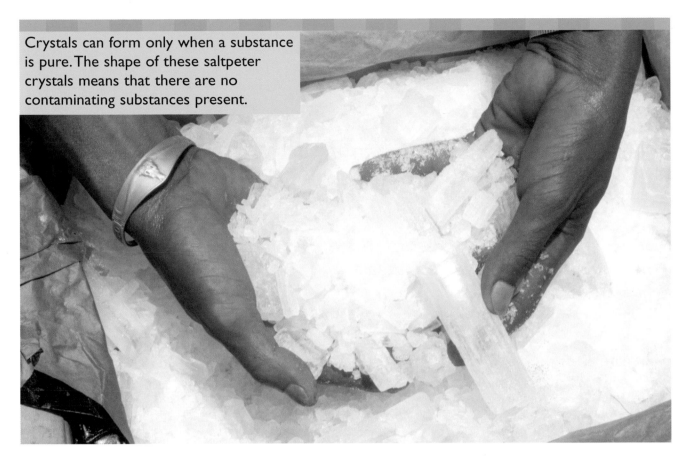

Crystals can form only when a substance is pure. The shape of these saltpeter crystals means that there are no contaminating substances present.

called *saltpeter*, or niter. Its chemical name is potassium nitrate. Saltpeter is one of the components in black powder. When saltpeter is heated, it releases oxygen. Any time you increase the oxygen to a flame, it will burn more brightly. (This is what happens when you gently fan a flame or blow on it.) Black powder explodes even when it is in a space that has no oxygen because oxygen is released by the saltpeter.

A Crystal Garden

Here's a fun way to see how crystals form on a surface in the same way that saltpeter does: In a small mixing bowl combine 3 tablespoons of salt, 3 tablespoons of water, 3 tablespoons of Mrs. Stewart's laundry bluing (see www. mrsstewart.com), and a teaspoon of ammonia.

Place a small sponge in a shallow dish. (It's fun to cut it into an interesting shape, like a fish.) Pour the liquid mixture over the sponge (1). Dot its surface with different colors of food coloring (2). Wait (3). Within two hours beautiful salt crystals will form on the surface and take on the colors of the food coloring (4). The sponge has acted like a wick. The solution has traveled up the sponge. Water has evaporated at the surface, leaving the colorful crystals behind.

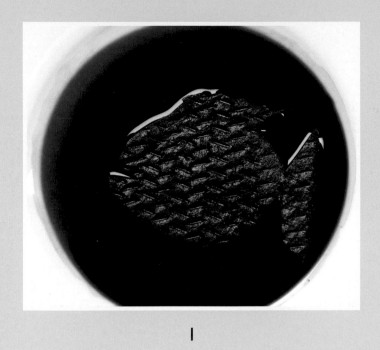

1

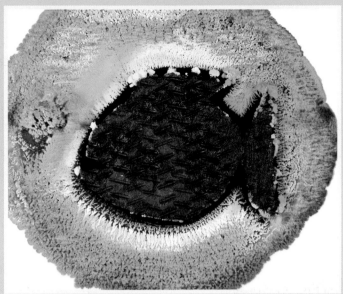

2

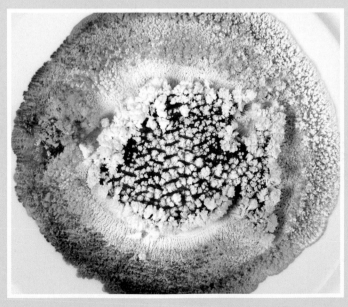

3

4

Saltpeter is formed by bacteria from the waste of animals and people. The bacteria extract nitrogen from the waste and combine it with oxygen. Potassium nitrate, the compound that forms, dissolves in water. As the water in the waste piles evaporates, flowery white crystals of saltpeter collect at the surface. Before modern plumbing, most outhouses had a drawer under the seat that could be emptied by a neighborhood night-soil collector. The night soil was put into tanks of water, where bacteria acted on it. Then the liquid was put into shallow trays. The water evaporated, leaving saltpeter crystals behind. Saltpeter can also be mined from mineral deposits.

Besides saltpeter, black powder also contains two fuels, sulfur and charcoal. The element sulfur is a yellow powder (see the facing photo). When it combines with oxygen, it burns with a blue flame and forms a smelly gas called sulfur dioxide. Charcoal is made by roasting wood so that a lot of impure chemicals are released as gas. What remains is almost pure carbon, also an element. When carbon combines with oxygen in the air it forms the gas carbon dioxide. These fuels are used in black powder because there is no hydrogen in any of the compounds. Thus no water vapor is produced. Water vapor would cool off the reaction and make it less explosive.

There are two reasons why black powder is so powerful: Its reaction is both rapid and very hot. The products of its combustion are two very hot gases: sulfur dioxide and carbon dioxide. Hot gases rapidly expand, creating a force that can propel a missile such as a cannonball, a bullet, or an aerial fireworks shell.

How a Match Works

Chemicals similar to black powder are used to make safety matches. The head of a match contains potassium chlorate, which gives up oxygen much as saltpeter does, and sulfur, which acts like a fuel. The striking surface contains powdered glass (to produce friction) and red phosphorus. When you strike the match, the heat produced by the friction makes the red phosphorus change into white phosphorus.

White phosphorus is so reactive that it immediately burns when it hits the oxygen in the air. The burning white phosphorus produces enough heat to cause the potassium chlorate to break down. This releases oxygen that combines with sulfur to produce a strong flame that can burn a wooden matchstick. Safety matches are safe because the phosphorus on the striking surface is separated from the chemicals on the match head.

The first friction match was invented in 1827. You could strike it on any surface, but there were lots of accidents. Safety matches were invented in 1855.

Cannons and Mortars

The first wars were probably fought by people throwing stones at one another. The invention of black powder made it possible to throw missiles farther than they had ever been thrown before. A kind of black powder was first used in China about A.D. 1000. It was packed in bamboo tubes in a kind of pipe bomb that exploded when lit. About 250 years later, it arrived in Europe, where people knew how to make balls of iron. Black powder was used to shoot iron balls out of a device known as a cannon. Basically, a cannon is a very strong metal tube that is closed at the back end and can be

The inside of a cannon barrel is "rifled," or grooved, to put a spin on the cannon ball as it explodes into the air. The rifling makes the cannon's aim more accurate.

tilted at different angles. Black powder was pushed to the back of the cannon through the open end, and a cannon ball was pushed up against it. When the black powder was ignited, through a tiny opening in the back of the cannon, the resulting explosion shot out the ball with enough force to carry it a significant distance.

A mortar, also called a "launch tube," is like a cannon but much shorter. Mortars are used to launch shells, explosive projectiles, into the air. In fact, aerial fireworks are called "shells." A fireworks shell can be shaped like a ball or a cylinder. It is placed into a mortar on top of some black powder in a plastic bag that is called the "lift charge." The amount of black powder determines the force of the lift. A small shell requires less of a lift charge than a larger, heavier shell. Most lift charges are designed to lift the shell about 100 feet (30 meters) into the air before it explodes.

These packets of black powder are ready to be inserted into shells.

Technicians load 4-inch (10-centimeter) cylindrical shells into a group of mortar tubes called batteries.

Rocket Science

When a missile is shot up into the air it makes an arc-shaped path, called its *trajectory*. There are several different factors that determine the trajectory of a shell. These include the force of the lift charge, the weight of the shell, and the angle of the launch tube. If you know the exact amount of each factor, you can predict with great accuracy what the trajectory of the shell will be. The kind of mathematics used for this calculation is algebra. Pyrotechnicians need to know when a shell is at the top of its arc in order to design an effective fireworks show. Today, computers do the calculations for a fireworks show.

Exploding Shells

Black powder isn't used only to send firework shells into the air. Each shell contains a black powder "bursting charge" that explodes to create the brilliant displays. The shells are cases made of hard paper similar to papier-mâché. Ball shells are easier to manufacture, but cylinders produce more spectacular displays. The bigger the shell, the bigger the size of the burst.

A master pyrotechnician hand rolls a casing.

Various shapes and sizes of aerial shells are organized for a computerized, musically choreographed fireworks show.

A 2-inch (5-centimeter) shell produces a burst that is 90 feet (27 meters) across, while an 8-inch (20-centimeter) shell produces a burst larger than three football fields across. The largest shells—up to 24 inches (0.6 meter) or more in diameter—are used only for special occasion, or for very expensive fireworks shows.

When shells are assembled, they are loaded with small combustible pellets called *stars*. Stars are made by mixing together black powder, chemicals that produce colors, and a binder to form a stiff dough. The dough is rolled out and cut into small cubes or rolled into pellets that

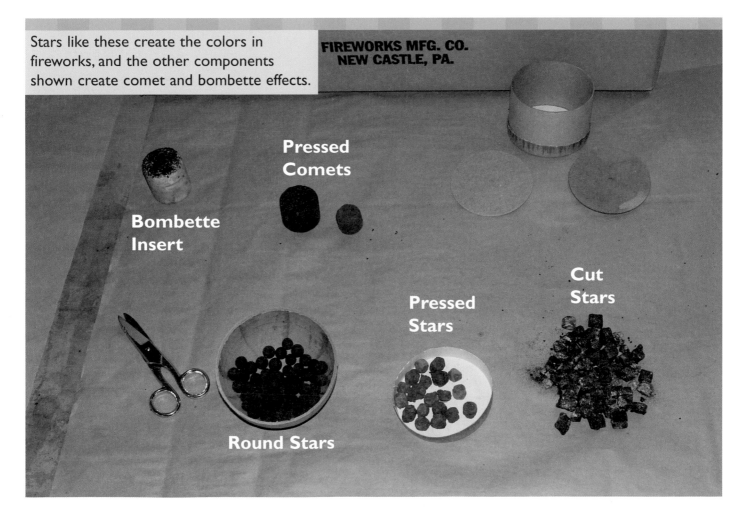

Stars like these create the colors in fireworks, and the other components shown create comet and bombette effects.

FIREWORKS MFG. CO.
NEW CASTLE, PA.

Pressed Comets

Bombette Insert

Cut Stars

Pressed Stars

Round Stars

This cup of pellets will be one break in a multistage effect.

are then set out to dry. Stars are packed in different patterns inside a shell, depending on the desired effect. When they explode, each star becomes a line of colored light against the sky.

In a ball shell, the stars surround the burst charge in the center. They explode to produce the chrysanthemum—a perfectly round pattern. The cylinder shells contain a random mixture of black-powder bursting charge and stars. They produce irregular splashes of light and color. A cylinder shell may also contain several compartments, or "breaks," that explode in a sequence. Each compartment has its own stars to produce a chain of effects. Often, the last compartment contains an aluminum powder mixture that gives off a loud bang at the end of the firework.

These are sparkling silvery starbursts with rising tails and gold streamers.

Special red and green multibreak crossettes keep coming from shells with many separate stages.

Tiny Fireworks

You can get a tiny form of firework from party stores. They are called "Party Poppers." If you try them, be sure to have an adult present.

The poppers are in the shape of a bottle with a string at the neck end. When you yank the string, you pull out the end, which has a rough material on it that produces friction. The friction provides the heat needed to ignite a tiny amount of black powder inside the bottle. The resulting "pop" causes enough of a blast of hot air to force out the cardboard bottom of the bottle. Inside are some tiny paper streamers that are also shot out from the force of the explosion. If you take one apart, you can see how it works. The smell after the explosion is from the sulfur in the black powder.

Colors in the Sky

Sunlight and other forms of white light contain all the colors of the rainbow. It's easy to remember the colors of a rainbow as ROY G. BIV, standing for red, orange, yellow, green, blue, indigo, violet. Light is a form of energy that travels in waves. Any wave, including a wave in the ocean, has a high point and a low point. The distance between the two high points or two low points is called the wavelength. Different colors have different wavelengths. Red has the longest, and violet has the shortest. The wavelengths of light you can see are extremely short, measured in sizes close to that of atoms.

When some substances are heated, they give off light. The color of a flame depends on the material burned. Carbon emits a yellow light; natural gas emits a blue light. Every glowing substance produces a distinctive pattern of wavelengths.

Different chemicals are added to fireworks to create different colors. Compounds containing the element strontium produce red fireworks. Strontium is always found as a compound in nature and is used mostly in fireworks. Sodium compounds, such as salt, produce a bright yellow-orange light. Barium compounds produce green. Aluminum and magnesium produce white light. Carbon and iron particles are used for a dimmer golden glow. The most difficult color to produce is blue. Copper will produce it but only if the temperature doesn't get too hot. If you see a rich blue color, know that you're watching a very professional show.

A pyrotechnician mixes chemicals formulated to create explosive and combustible materials. His face is protected to keep the chemicals from entering the eyes, nose, or mouth.

A Flame Test

Do a flame test to see some of the colors certain metals give off when their compounds are heated. Since you will be using a flame, do this with an adult.

You will need a metal (not copper) paper clip, a pencil with an eraser, a gas flame from a stove, rubbing alcohol, and compounds containing metals such as salt (sodium), salt substitute (potassium), and minerals containing calcium, such as calcium tablets. Straighten the paper clip and make a loop at the end. Stick the other end into the

eraser of a pencil. Dip the loop into the alcohol. Then dip it into a chemical such as salt. Have an adult stick the loop into a flame on the stove. The color the flame turns depends on the metal: Sodium gives off yellow, calcium orange, potassium violet.

Timing the Explosions

How can an explosive be lighted while allowing time to get out of the way before it goes off? A fuse is the answer. The first fuse was probably a trail of black powder wrapped in a tube of tissue paper. When the end was lit, it burned along its length until it reached the firework. Fuses have also been made of string with black powder embedded in it. But today's fuses are an electrical "match." A spark coming out of the end of a wire ignites the firework. In a modern aerial shell, a signal is given from a computer to send an electric current through a "leader" wire. The leader sparks to ignite a fuse that is connected to both the lift charge and a time-delay fuse inside the shell. The time-delay fuse will ignite the blast charge when the shell is at its highest point in the sky. If there are breaks in a shell, the time-delay fuse sets off one break and continues burning to set off the next one. The lengths of the fuses, their burning time, the size of the lift charge, and the weight of the shell are all carefully calculated. Precision measurements make sure that the shells explode at the right time and the right place in the sky.

Twenty minutes of a spectacular fireworks show may have more than two hundred aerial shells and more than eighty "illuminations" of fireworks, like Roman candles, that burn from a stage or barge. The sequence of explosions has been carefully planned and choreographed to

Pyrotechnicians load aerial shells for a choreographed musical production, which will be fired by a laptop computer connected to electrical "matches."

music. The mortars are placed in groups called "batteries." A battery of mortars in warfare increased the firepower of the weapons. The word "battery" stuck and is used for a group of mortars in a fireworks show. Each mortar is loaded with shells, and leaders are attached to the fuses. The leaders are connected to wires that lead to an electrical firing panel. At one time, pyrotechnicians pushed the buttons on firing panels to set off the fireworks. But today a computer program does the job.

The electrical matches, or squib wires, have been inserted into the shells' leader fuses. This pyrotechnician is attaching the squib wires to an electrical field box, which will transfer the electrical impulses from the electrical firing board to the squibs, igniting the shells.

This electrical firing board is one of several methods used to ignite fireworks displays.

Avoiding Nasty Surprises

People who work with fireworks know that you can't be too careful. There are two famous Italian-American families, the Gruccis and the Zambellis, who have made fireworks their business. They know, better than most, that fireworks are dangerous. Injuries from fireworks cost more than $100 million a year. Boys between the ages of 10 and 14 have the most fireworks-related injuries. And almost a third of the injuries come from illegal fireworks. Even sparklers burn at temperatures more than 1000°F (538°C).

Here are some warnings for pyrotechnicians that are often included in their training manuals:

There are old pyros and there are bold pyros, but there are almost no old, bold pyros.

It's better to have a fuse a foot too long than an inch too short.

The Grucci family of Long Island, New York, learned these lessons the hard way. In 1983 an accidental explosion at their building complex killed two members of their family. They have put a great deal of thought into finding ways to prevent this from ever happening again. Employees who make the shells work in many small buildings, not one big plant. That way, if there is an unfortunate accident, the explosion is not as large as it would be if all the explosives were under one roof.

The greatest fear is that a spark could set things off, and static electricity is one frightening source of sparks. Workers wear only cotton

This is one of the Zambelli's work buildings. If there was an explosion, it would come out a door on the left. The concrete wall protects the force of the blast from damaging other buildings.

clothes, including underwear. (Synthetics like nylon or polyester collect a static charge.) When employees enter a workroom, they touch a copper plate at the door. Any static buildup will be conducted by the copper into the ground. There are no electrical wires going into a workroom. And if a worker wants to listen to music, the radio sits outside near a window.

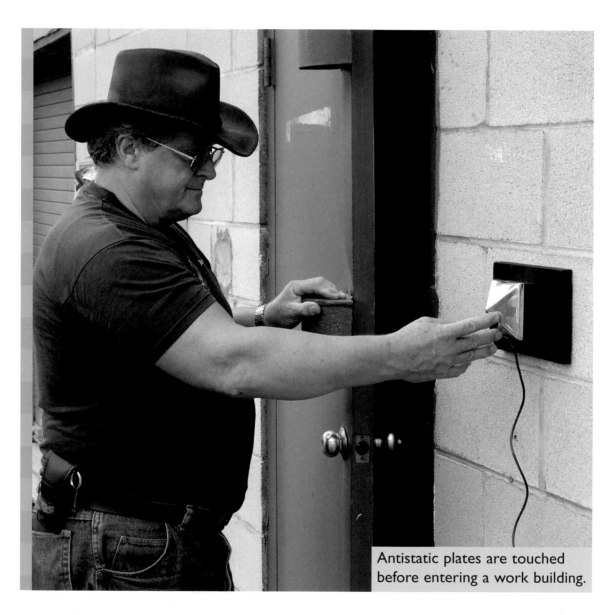

Antistatic plates are touched before entering a work building.

Finished shells are stored in many 40-foot (12-meter) containers that look like the trailer part of a tractor-trailer truck. The door to a container never faces the door of another container. That way, if one explodes, it won't set off any of the others.

There are strict rules about becoming a professional pyrotechnician. You must be at least twenty-one years old, and you must attend about fifteen hours of training. You must hold a license, and you will not get a chance to run your own show until you have been an assistant on at least ten shows.

For experts like the Gruccis and the Zambellis, satisfaction comes from designing a fabulous show and making it work exactly as planned. Understanding how these incredible displays are produced makes them even more awesome. Now you can be the expert and share the knowledge with your friends as they ooh and ahh the next time you go to a fireworks show.

Forty-foot steel magazine containers are used to store fireworks at a manufacturing plant.

Chrysanthemums with Titanium Salute

Chrysanthemums with Rising Tails

Gold Brocaded Kamuro

Grapes in the Vineyard

Green Chrysanthemum with Red Pistil

Rainbow Candles

Silver Willow

Spider Web

Double Rings Shells

The Names of Fireworks

Battle in the clouds: A number of salutes timed to explode in series.

Chrysanthemums: Stars that burst from a central core with trails.

Diadems: Cardboard-like strips or disks that burst out of a shell and flicker brightly as they fall in a cluster.

Hummers: Small tubes that spin and create screaming or humming sounds.

Patterns: Stars that explode in a shape such as a 5-pointed star or a heart.

Peonies: Stars that burst from a central core without trails.

Salutes: Loud, concussive booms.

Set pieces: Wooden frameworks set on the ground and studded with lances. When ignited they become a picture in colored fire.

Strobes: Clusters of flashing silvery lights that appear to float slowly to the ground.

Tourbillions: Ignited shells that spin like tops as they rise.

Weeping willows: Star colors that burn with an amber hue as they fall, outlining the dropping branches of the willow.

Whistles: Noise-making tubes that dart across the sky, giving off shrieks as the shell bursts.

Key Words

To find out more, use your favorite search engine to look up the following terms on the Internet.

chemical elements

chemistry

combustion

compound

fireworks

Grucci

mortar

physics

pyrotechnician

saltpeter

trajectory

wavelengths

Zambelli

Index

Page numbers in *italics* refer to illustrations.

About the Author

Ever since *Science Experiments You Can Eat*, Vicki Cobb has delighted two generations with her scientific and playful look at the world. In the "Where's the Science Here?" series she pays attention to areas kids know are FUN. Take fireworks, for instance. Her favorite kind of July fourth is an old-fashioned small town celebration on the shores of the Hudson River. Now she finally knows the names of the spectacular effects she enjoys. At other times of the year she's writing books and talking with kids and teachers. Visit Vicki at: www.vickicobb.com.

About the Photographer

Michael Gold is a commercial photographer who has worked on assignment for some of the most exciting accounts, including *The New York Times*, *Fortune*, *Esquire*, *American Express*, *BMW*, *Mobil*, *Opera News*, and many more. His work includes food, internationally known celebrities, advertising, products, fashion, and corporate photography. He has had nine one-man exhibitions, portfolios published in *Popular Photography Magazine* and *Camera 35 Magazine*, and is included in *LIFE*'s first humor anthology, "LIFE Smiles Back" and "Who Needs Parks?"